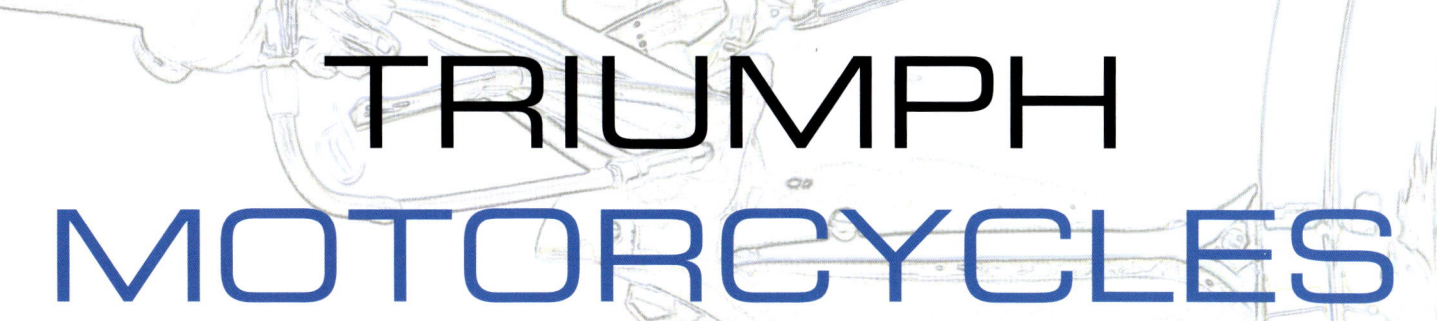

TRIUMPH
MOTORCYCLES
From Speed Twin to Bonneville

By Timothy Remus

Published by:
Wolfgang Publications Inc.
217 Second Street North
Stillwater, MN 55082
http://www.wolfpub.com

Legals

First published in 2005 by Wolfgang Publications Inc.
217 Second Street North, Stillwater MN 55082

© Timothy Remus, 2005

All rights reserved. With the exception of quoting brief passages for the purposes of review no part of this publication may be reproduced without prior written permission from the publisher.

The information in this book is true and complete to the best of our knowledge. All recommendations are made without any guarantee on the part of the author or publisher, who also disclaim any liability incurred in connection with the use of this data or specific details.

We recognize that some words, model names and designations, for example, mentioned herein are the property of the trademark holder. We use them for identification purposes only. This is not an official publication.

ISBN number: 1-929133-21-9

Printed in China

TRIUMPH MOTORCYCLES
From Speed Twin to Bonneville

Chapter One
    In The Beginning 6

Chapter Two
    Generator Power 28

Chapter Three
    650cc & 2 Carbs 66

Chapter Four
    Two Magic Letters 120

Chapter Five
    A Sad Ending 128

Chapter Six
    New/Old Triumphs 138

The Catalog 142

Acknowledgements

The images that makeup this book are the result of over ten years experience photographing classic Triumph Motorcycles. It was Bobby Sullivan who started the ball rolling with a suggestion that we photograph some of his Bonneville and TR6 bikes during Daytona Bike week, way back before I had any gray hair. That friendship and shared appreciation for certain old English motorcycles spawned the Classic Triumph Calendar. If Bobby can be counted on for anything, it's his endless energy and wild ideas, which led to motorcycles perched in impossible situations, all to get a good shot. My thanks then start with Bobby and all the various and changing members of the Sullivan crew.

Publication of the calendar eventually became my responsibility, which in turn led me to Randy Baxter in Iowa. Without Randy Baxter and Baxter Cycle there would be no Classic Triumph Calendar, and no hardcover book. My summers have come to include two road trips to Iowa to photograph the ever changing collection of vintage Triumphs that pass through Randy's facility in Marne, Iowa. My "thank yous" continue with the crew from Baxter Cycle, (especially Art) for all their help washing, polishing and moving motorcycles.

Ultimately the best thing about motorcycles isn't the bikes themselves but rather the people they seem to attract. Those summer trips always include a stop outside DesMoines to photograph three or four bikes owned and restored by Denny Narland. Denny not only makes his own bikes available, and helps with early morning photo shoots, he convinces other Triumph owners like Denny Beckman to make their bikes available as well.

For photos I couldn't capture myself, I thank Jeff Hackett, a photographer with his own unique view on the world and a special knack for isolating motorcycles in space. For the wonderful design of this book I tip my hat to our in-house designer, Jacki Mitchell. And for ensuring we pay the printer (and Jacki) on time, we all thank Krista Leary. For moral and immoral support, and putting up with my frequent absences while I go first to a book-expo and then a last-minute photo shoot I thank my lovely and talented wife Mary Lanz.

Introduction

Though my obsession with Triumphs started in high school, it was only recently that I actually became a Triumph owner. Between then and now I've satisfied my fascination first by reading books, then by authoring my own. Those earlier writing efforts attempted to document specific details, like the first year some obscure detail was used on a Triumph.

This book skips all the minutiae true Triumph nuts pour over and carry around in their head. The book you hold is more fun than that. This a photo book - period. The introductory comments in each chapter and the captions provide a light overview of Triumph's history. Just enough to place the bikes in a context and help explain the evolution of Triumph Motorcycles.

Team Triumph, Edward Turner and his team of engineers and stylists, designed machines that were both very fast for their day and very beautiful. Luckily the idea of beauty is a less fleeting than the idea of fast. The soft organic curves, the harmonies of the various castings and covers, the choice of colors and the way the colors contrast with the abundant chrome and polished aluminum, all make for a very good looking motorcycle, whether viewed thirty years ago or today.

Call this book a visual celebration. A party on paper. A chance to sit back at the buffet table and revel in the magic that is Triumph.

In The Beginning

The history of the Triumph Motorcycle Company can be distilled down to only two words: Speed Twin. Though the company made both singles and triples it was, and is, the twins that put Triumph on the map. While the rest of the two-wheeled world made do with staid thumper singles, heavy V-twins and even a few big four-cylinder designs, Triumph gave them the perfect alternative.

Designed by Edward Turner and introduced to the world in 1937, the very new bike gave riders of the day the speed and flexibility of a twin in a package no bigger than a single. In fact, the Speed Twin, with 26 horsepower actually weighed less than the biggest of the Triumph singles.

Designed to be fast, simple and light, the new twin used a 360 degree crankshaft that kept the two pistons moving in tandem but on opposite strokes. That is, while one piston was compressing, one was exhausting, when one sped downhill on the power stroke the other came down at the same rate on the intake stroke.

Weight reduction came about through both the use of lightweight materials and clever design. The designers used aluminum for the cases, and then specified cast iron for the cylinders - which they cast in one piece. By specifying the right aluminum alloy, Edward Turner was able to use connecting rods that ran directly on the crankshaft without any bearing inserts. The true strength of a great design lies not just in its acceptance when introduced but in its success over time.

Pointing the way to the future, the Speed Twin used a typical Triumph girder fork (borrowed from the biggest single cylinder bike), a large diameter headlight with bright chrome housing, and a speedometer driven off the front wheel. Later model Speed Twins with the girder design used a check spring on either side of the fork.

In the beginning God made the Speed Twin. Lighter, faster and better looking than nearly anything else on the planet at that time, the bike was a sensation then and still a damned fine piece of work sixty-some years later.

Built to roll down the highway, the first Speed twin used a 20 inch front tire and a 19 in the rear. Rims are chrome plated with a red stipe running down the center. Though the brakes look inadequate today, they were standard-issue for 1938.

Early cylinder heads used external oil drain-back tubes. Many of the parts on the first Speed Twins were borrowed directly from Triumph's biggest single-cylinder machine.

Ammeter and oil pressure gauge were set in the tank-top, as was the practice with many Triumph models. The Speed Twin design was essentially one-half of the Aerial Square Four, another Edward Turner design. A very compact design, both the cylinders and head were one single casting. Though later heads would be made from aluminum, the Speed Twin used cast iron for both the cylinder and the head castings. Note the number of studs holding the cast iron cylinder to the alloy cases.

Specification – "De Luxe" Models

ENGINE: Models 5H, 3H, 2H and 2HC. Single port O.H.V. deeply finned cylinder barrel and head. Piston of special low expansion aluminium alloy. All-gear drive to magdyno. Coil ignition to Model 2HC.
Models 6S: 5S and 3S: Side valve units with quickly detachable cylinder head. Latest non-pinking internal formation to combustion chamber. All-gear drive to magdyno.
CONNECTING RODS: "H" Section. Nickel chrome stamping combining great strength with lightness. Double roller bearing big end.
CRANKCASE: Aluminium alloy heavily webbed internally, magneto platform cast integrally with timing case.
CRANKSHAFT ASSEMBLY: Crankshaft of substantial size mounted on large diameter ball bearings. High tensile forged steel flywheels.
VALVE SPRINGS: Aero quality. Duplex springs.
LUBRICATION: Full dry sump system. Valve gear fully enclosed and automatically lubricated. Plunger type oil pumps with positive feed to big end. Oil gauge in instrument panel.
CARBURETTER: Amal. Triumph patented quick-action twist-grip throttle control.
PETROL TANK: All-steel welded, combining shapely streamline contour with large capacity. Die-cast all-metal permanent nameplate. Rubber mounted instrument panel of moulded construction carrying oil gauge, ammeter, switch and new, exclusive external panel light. Petrol capacities:
Models: 6S and 5H. 3½ gallons.
Models: 5S, 3H, 3S, 2H and 2HC...3 gallons.
OIL TANK: All-steel welded with accessible filters, drain plug and separate vent. Capacity: ⅝ gallon, all models.
FRAME: Full cradle type, with single large diameter front down tube. Soundly constructed from tubes of finest alloy steel. Great strength and torsional stiffness with low weight. Excellent weight distribution is afforded and the roadholding at speed is of the highest standard.
FRONT FORKS: Taper tube girder type, light but strong, with hand adjustable rebound dampers on lower bridge.
GEARBOX: Four-speed all-Triumph design and manufacture. Gears and shafts of hardened nickel chrome steel, finished to the highest possible standards of accuracy and precision. Patented positive stop foot-change—fully enclosed. Short pedal movement ensuring delicacy of control with completely positive action. Large diameter multi-plate clutch with enclosed operating gear; light in operation with smooth engagement; accessible adjustment. Lubrication by engine oil.
TRANSMISSION: Primary chain running in polished cast aluminium oil-bath chain case of streamline design. Rear chain positively lubricated by feed from primary chain case.
BRAKES: Triumph 7" diameter brakes with extra wide shoes. New brake lining material giving powerful and smooth braking with long life. Finger adjustment. Front brake adjustment accessible from saddle.
SADDLE: De Luxe soft top type, adjustable for height.
HANDLEBAR: New Triumph design with modified bend, resiliently mounted, eliminating fatigue and shocks and giving great riding comfort and controllability. Control levers grouped and adjustable to suit individual requirements. Long type brake and clutch levers.
MUDGUARDS: "D" Section, and of adequate width. Detachable tail piece to rear guard to facilitate rear wheel removal. Streamline section stays. Valanced guards available on Models 5H, 6S; at extra charge.
WHEELS AND TYRES: Latest Triumph wheels of improved design. Dunlop Tyres.
Models 6S, 5H, 5S, 3H and 3S: Front and rear...3.25".
Models 2H and 2HC: Front 26" x 3". Rear 26" x 3.25".
TOOLBOX: Shapely all-steel construction, of large capacity, with rubber sealed lid ensuring protection against water. Complete set of good quality tools and grease gun.
FINISH AND EQUIPMENT: Petrol Tank in chromium plate, panelled in Black, lined out in Ivory. New design of knee grips combining maximum comfort with security at high speeds. All aluminium parts highly polished. Frame, forks and mudguards finished in three coats of the highest quality black enamel. Triple Ivory lines to mudguard centres. Wheel rims chromium plated, black centres, lined out in Ivory. All bolts Cadmium plated. Lucas 6-volt magdyno lighting with voltage control, large diameter Lucas headlamp and electric horn. Lucas coil ignition equipment on 2HC Model. All aluminium parts smooth and highly polished, and both chromium plate and enamel of the highest quality.

Illustrations of "De Luxe" Range will be found on pages 11 to 16. Exclusive Features on page 19. Technical Information on page 20.

The brochure image dates from 1939, note how the the frame, fork, oil and gas tank, and tool box, are all the same (or very close) to the parts used on the Speed Twin. Part of the genius of the Speed Twin was the fact that so many existing Triumph parts could be used without modification.

TRIUMPH DE LUXE 5H

500 c.c. O.H.V.

PRICE: £62

Fully equipped with Lucas Magdyno lighting and Electric Horn.

A Smith illuminated Chronometric Trip Speedometer (80 m.p.h.) will be supplied unless otherwise ordered, £2-10-0 extra.

Even a good design can be improved. In the case of the Speed Twin, that improvement came in 1939 with the introduction of the Tiger 100. With increased compression and a ported head for better breathing the new model would top out at over 100 miles per hour, thus the name. A silver paint job with pinstripes (the pinstripes painted by hand of course) helped to separate Triumph's first hot rod twin from the more common Speed Twin.

The original Speed Twin proved to be a good enough design that very little change or improvement was called for during the first few years of production. This model from 1939 uses the same 500cc engine and four speed transmission used on the introductory model. Though the first bikes used six studs to secure the cylinders to the engine cases, later "eight stud" bikes eliminated one of the few weaknesses in Turner's original design.

Triumphs have always had a certain style and attitude. Even the early Speed Twins used abundant chrome at a time when chrome was expensive and much less common than it is today. The large diameter headlight with its sparkling housing, and the gauges contained in the tank, are all part of what helped the Triumphs stand out from other bikes of the period.

As seen in the factory's brochure, the Speed Twin was not only a great motorcycle, it also served as the foundation for an entire company that would go on to prosper for over forty years. Not to mention all the other vertical twin copies from all the other English firms.

Not unlike some of the current Harley-Davidsons, the T100 actually differed very little from the sibling Speed Twin, yet was accepted without question as a completely different model. More horsepower, different paint and trick "cocktail shaker" mufflers with the removable end caps identify this bike as a 1940 T100. Photo Mark Mitchell

Production of motorcycles at the Coventry plant stopped abruptly on November 14, 1940 when a German air raid leveled the facility and much of the surrounding city. The bike seen here is one of the last bikes produced at that plant. Though there were no deaths among the night shift, the raid destroyed all the tooling needed to produce a new 350cc military bike, the 3TW. After the raid Triumph moved the manufacturing facilities into temporary quarters in nearby Warwick.

The early war years was no time to be improving existing products. This 1940 Speed Twin uses the signature Girder fork, as Triumph didn't introduce hydraulic forks until after the war. Likewise the polished primary cover which hides the standard single row primary chain connecting the engine with the four-speed transmission. By 1942 Triumph was producing very basic 3HW singles for the military, based on a civilian model, at a brand new plant located outside Coventry, near the village of Meriden.

The Speed Twin changed very little from it's introduction in 1937 to the start of the war. Owned by Baxter Cycle, this prime example from 1940 uses the same 360 degree crankshaft supported by ball bearings at either end, specified for the first Speed Twin. And like its predecessors, this bike uses two identical camshafts, the exhaust in front and the intake at the rear.

Q&A Denny Narland

Pulling in the long driveway at Denny and Kathleen Narland's house near DesMoines you pass under a canopy of trees until eventually you come to a cement slab. The building on the left appears to be a free-standing garage, on the right sits a good sized house with a whole raft of garage doors facing the driveway.

Closer inspection reveals the building on the left to be a combination shop and office for Denny, with an attached smaller office for his wife Kathleen. Once in the door there are Triumphs everywhere you look. The shop/office includes a lift and workbench for Denny's current project. But all the floor space is occupied by Triumphs as well, including part of Kathleen's office and a good chunk of the garage space on the other side of the driveway. What started with the purchase of one Bonneville more than thirty years ago has turned into a full blown hobby. Denny calls it a "controllable hobby," though after looking at the considerable amount of space taken up by Triumphs of every description and state of repair, you have to wonder about how controllable the hobby really is.

Q & A: Denny Narland

How did you get hooked up with Triumphs?
I bought my first one in 1969, well after I graduated from high school. My friends had them and I always liked Triumphs. I had my first bike at 14, a Simplex service cycle with belt drive centrifugal clutch. And I had other bikes too, a 1948 Panhead, a 1970 Sportster. I bought the 1969 Triumph one year before going into the service, then after finishing basic training I took the Triumph with me to Ft. Hood, Texas.

You were involved in the motorcycle business, tell us about that?
It was after I got out of the service, from 1971 through 1976. The name of the shop was Dan's Cycle Bones because it started as a salvage place but we had all kinds of parts for all the major brands. Dan Brewer was a Triumph rider, amidst all the Harley riders in the area. He was the first guy locally to put a 16 inch wheel on a Triumph. He rode Triumphs with other mostly Harley riders.

As the shop evolved into Midwest Cycle Supply we got a Norton dealership. Then we had opportunities to pick up other brands, eventually we handled Moto Guzzi, Husqvarna and Ducati. We tried to get Harley and Kawi too but never did. The movie, On Any Sunday, really hit home because we raced every Saturday night and Sunday.

Q&A Denny Narland

After I left the shop in 1979 I had chance to buy another 1969 Triumph, and from that another and another. I'd been through the cars, Corvettes and all the rest and then realized I could go faster on a twelve hundred bike than a five thousand dollar car. Triumph was the fastest production 650 of the period. The Nortons were a little faster but they were 750s. When I first saw Triumphs it became something I sought to have. At one point it was the thrill of a lifetime to own one.

How many do you own now?
I would have to guess, probably about 20 nice complete bikes and another 20 crying to be restored. Seldom do I sell one, though I have sold some over the years. Often the ones I sell are later model bikes or project bikes that I intend to fix up and sell. Denny Beckman talked me out of a number of them, probably seven or eight bikes.

Do you a lot of restoring and repairs for other people?
There are only three I can think of that I restored for other people. The way I see it, this is a controllable hobby. I like to see them in progress. I like to see them when they're finished. They sit patiently waiting for me to come back to them.

What are your favorite years?
1969 is my favorite, because it was questionably one of the best years, I wanted a 1969 when they were new, and I would still like to have the whole model year range. I do have them all but I still have to restore one.

My area of interest is 1966 through 1970. Seldom do I go before that due to lack of knowledge and seldom after that because of the BSA influence. And I can barely keep up with the years I do. In my years of interest, I got involved with them when you could still get a nicer original bike. I have at least 5 bikes where I'm the second or third owner.

Like the story of the TT Special, it was one of two bikes that three friends (who I don't know) bought to trail ride. It went from Michigan to Florida, the guy died, his wife called the two friends. They took it to Daytona, then it went to Maine and came back to the Midwest again. It's just amazing that I became the owner of that bike.

Finding the information and history of the bike is fun too. Like that TT, it had been ridden, but not abused like so many of them were. This Bonnie over here, the second 1969 I acquired, I got it from a guy who's Father bought it new. When I purchased the bike it had a 1969 Kansas license plate. I couldn't talk him out of the license plate but I did talk him out of the bike. Experiences like that are so few and far between now. It's become hard to find bikes that belong to the original owner or a relative of the original owner.

How much do you ride these bikes, and how many run on a daily basis?
At least three run at any given time. 'Cause if you ride 'em, and I do, then two can be broken down at any time and you've still got one to ride.

After the war many materials needed for manufacturing were rationed by the Ministry of Supply. Allocations depended on how much product you exported. Luckily for Edward Turner and Triumph, many returning GIs were both familiar with English products and ready to buy a new motorcycle. When the war ended Triumph simply revived the models sold earlier. Thus the gas tank and signature gauges are the same as what was used before the war.

The end of hostilities meant a conversion from military single cylinder bikes painted olive green to civilian bikes with two cylinders and abundant chrome plate. Framed by a 1944 T-6 trainer, this hot T100 dates from 1946. Still displacing 500cc, the biggest change to both Speed Twins and T100s produced after the war can be seen at the very front of the bike - the hydraulic fork installed in place of the classic Girder design used for so many years.

The design of both the Speed Twin and T100 bikes make good use of chrome and paint. Though there's a great deal of chrome on this bike, especially considering the period, it's still an accent rather than the main effect. Even the rims, measuring 20 inches in front and 19 in the rear are chrome plated with painted centers. Think about how good those gas tanks had to be before they could be chrome plated without showing any ripples, high or low spots, or evidence of metal finishing.

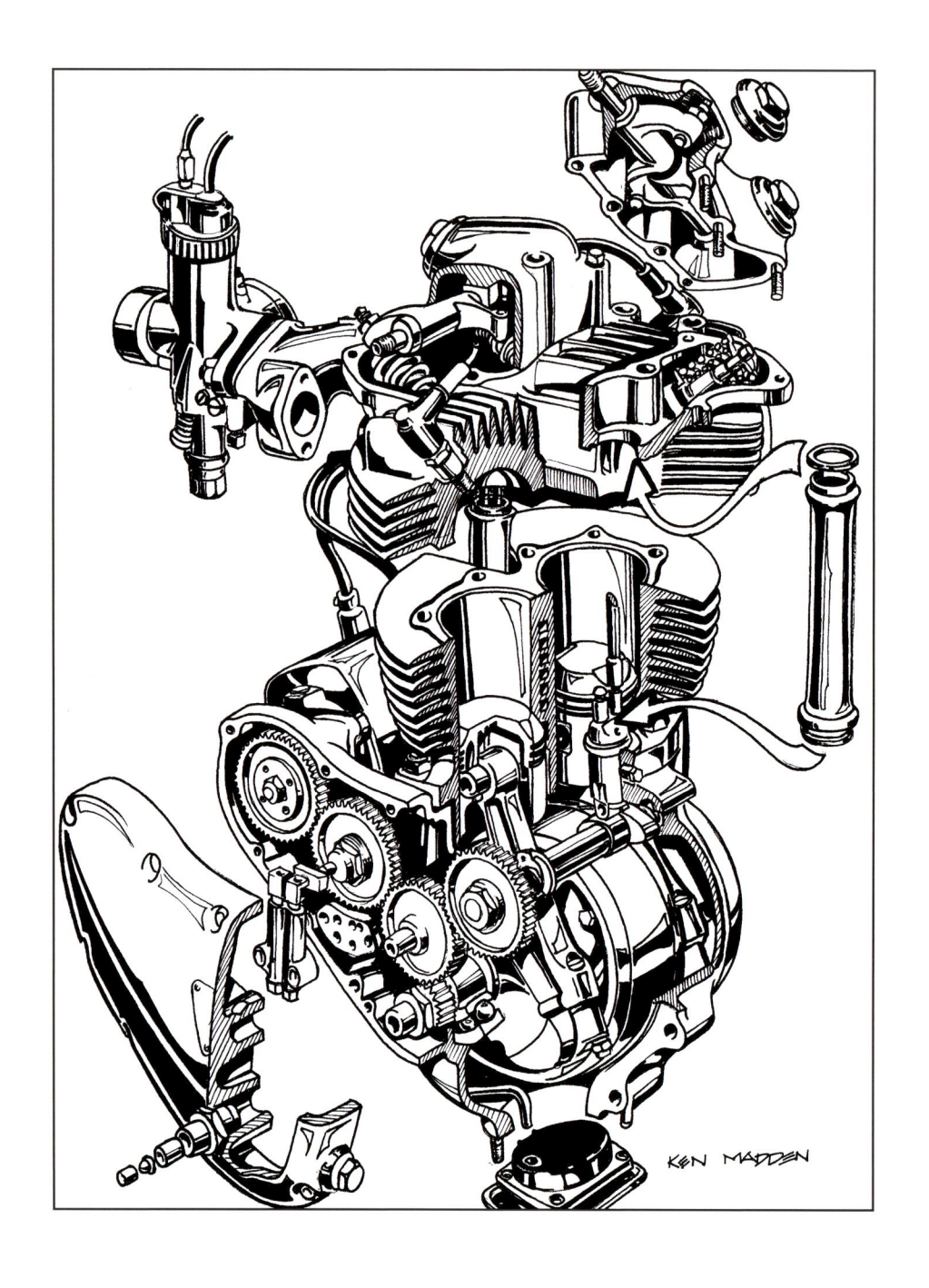

Whether you're talking about the first Speed Twin, a later T110 or even a Bonneville, Triumph motorcycles are all about the engine. Though criticized later for their vibration, the parallel twin as developed by Edward Turner was a brilliant design. One that gave riders of the time more power and less vibration than the big singles of the period with little or no increase in weight. Edward Turner wasn't the first to market a parallel twin, but he was the first one to make a parallel twin both powerful and light, the first one who got it right.

Closely related to the 3TW military design, the 350cc 3T was never produced in large numbers, before or after the war. Anemic performance, even for a 350, meant poor sales and an early death in 1950.

Generator Power

The Triumph Motorcycle Company contributed to the war effort in various ways. In addition to the 350cc 3HW, and the later TRW, Triumph produced a 6KW generator designed to recharge the batteries of the big bombers while in the air. The powerplant half of the generators was designed around the then-current 500cc engine. To keep the engine light however, the top end was cast entirely from aluminum. To make the whole thing fit in a rectangular sheet-metal housing the cylinders were squared off.

Freddie Clarke was the first to mate a 'generator' top end with a standard T100 bottom end. Head of Triumph's Engineering Department, Eddie won the 1946 Senior Manx Grand Prix on just such a machine. The experimental grand prix machine proved successful enough to spawn nearly 200 official factory duplicates, one of which landed in the hands of American Rod Coates who used the machine to win the 1950 Daytona 100 mile race.

The same basic recipe was used to power a new model known as the Trophy. Named after the Team Trophy won by Triumph riders competing in the International Six Day Trials, the new off-road racer used a short wheelbase frame, a "generator" engine and lower gearing to become what Lindsay Brooke calls "the first true dual-purpose twin."

Pent up demand for faster and faster motorcycles, especially in the US, meant the new TR5 was only the first of a long line of faster and faster motorcycles to come out of Meriden.

Modified for racing by none other than Dick Mann, this early TR5 with the legendary generator top end is the property of Bobby Sullivan. Note the rectangular shape of the barrels and the un-drilled bosses, originally used to attach the shielding when the engines were used in generator duty. Unlike the Grand Prix models, the TR5 bikes used milder camshafts and lower compression to create an engine with abundant torque at nearly any RPM.

The TR5 used a 20 inch front rim and 19 inch rear fitter with 3.10 and 4.00 Dunlop tires. Though the original bikes came with full aluminum fenders, these have been "abbreviated" for full competition duty. Though these early TR5 bikes with their generator engines were highly competitive, a major upgrade was in the works.

A more pristine TR5 as delivered by the factory in 1949. The siamesed exhaust would be a standard item on certain Trophy models for years to come. Abundant torque and light weight combined in a short-wheelbase frame made for a nimble and unstoppable off road machine. The Trophy bikes came by their model name honestly. Jeff Hackett photo

By 1948 the Tiger 100 was still the hot rod street bike of the Triumph line and a very popular ride on both sides of the Atlantic Ocean. This machine, restored by Baxter Cycle, is painted silver sheen and accented with black pinstripes. Large chrome headlight housing and cocktail-shaker mufflers, used on earlier T100s, were long gone by 1948. And the signature gas tank, with built in gauges and service light, would soon be gone as well.

Post-war changes to the Triumph line included both upgrades to existing models and the introduction of new models like the Trophy. By the late 1940s Triumph was busy trying to meet the demand for existing models, and planning new models with more modern styling and more power. For 1948 the T100 used the same fork, engine and four-speed transmission seen on the 1946 Tigers. Which is not to say the T100 was an old design. In the hands of both amateurs and professionals, the bikes were consistent winners on both dirt and asphalt.

Specification
498 c.c. O.H.V. "TIGER 100"

The TRIUMPH "TIGER 100" is an ultra high performance sports machine with a specially tuned engine capable of completely satisfying the desires of all who wish to travel fast and far. At the same time it retains that flexibility and smoothness which make it a very pleasant motor cycle when high speeds are not desired. Finished in the well-known "Tiger" silver sheen, black and chromium.

TRANSMISSION. Primary chain in polished cast aluminium oil bath case. Rear chain positively lubricated and protected on top and bottom runs. FOUR-SPEED GEARBOX. Triumph patented design and manufacture. Gears and shafts of hardened nickel chrome steel. Patented positive stop foot-change, fully enclosed. Large diameter multiplate clutch, light in operation and with accessible adjustment. Gear ratios (solo) 5.0, 6.0, 8.65 and 12.7 to 1.

PETROL TANK. All-steel welded streamline design of 4 galls. capacity. Flush fitting, rubber mounted instrument panel incorporating ammeter, oil gauge, light switch and inspection lamp. Quick opening filler cap and die-cast metal nameplate.

OIL TANK. Extra large capacity (1 gallon) all-steel welded with accessible filters, drain plug, separate vent and quick release filler cap.

FRAME. Brazed full cradle type, with tubes of finest alloy steel. Large diameter tapered front down tube.

FRONT FORK. The famous Triumph Telescopic pattern with large movement, hydraulically damped, automatic lubrication. No adjustments necessary.

BRAKES. Triumph design with large braking area and finest quality lining materials. Finger adjustment front and rear.

HANDLEBAR. Special Triumph design. Fully adjustable chromium plated control levers.

MUDGUARDS. Wide "D" section with streamline stays. Detachable rear guard. Triumph patented front number plate and streamline rear plate with centrally mounted lamp.

WHEELS AND TYRES. Triumph design wheels. Dunlop tyres, front 19 × 3.25 (ribbed), rear 19 × 3.50 (studded).

TOOLBOX. All-steel large capacity with weatherproof protection. Complete set of good quality tools and greasegun.

EQUIPMENT. Lucas 6 volt dynamo lighting set with voltage control, large diameter head lamp and electric horn. Special Triumph design kneegrips, adjustable de Luxe saddle and downswept exhaust pipes with large capacity silencers. Smith 120 m.p.h. speedometer with R.P.M. scale and internal illumination.

FINISH. Petrol tank chromium plated with silver sheen panels lined out in blue. Mudguards in silver sheen with black central stripe. Wheel rims chromium plated with rim centres in silver sheen lined blue.

SPRING WHEEL. Available as an extra. For full details see Page One.

PROP STAND. Extra. For details see Page Eight.

ENGINE SPECIFICATION

Vertical twin cylinder with gear driven double high camshafts. Bore 63 mm. Stroke 80 mm. Capacity 498 c.c. Cylinder heads, ports and all moving parts highly polished. Special high compression pistons of silicon low expansion alloy. Totally enclosed and positively lubricated valve gear, highly polished rocker boxes and pushrod tubes. Duplex aero quality valve springs. High tensile aluminium alloy crankcases. "H" section connecting rods in RR 56 hiduminium alloy with patented plain big-ends. Patented crankshaft mounted on heavy duty ball bearings with central flywheel. Full dry sump lubrication incorporating plunger type pump with positive feed to big-ends and valve gear. Amal large bore carburetter with patented Triumph quick action twist grip. Automatic advance magneto and separate dynamo all-gear driven.

TRIUMPH TIGER '100'
PAT. NOS. 475860, 474963, 482024, 469635.

34

Foreword 1948

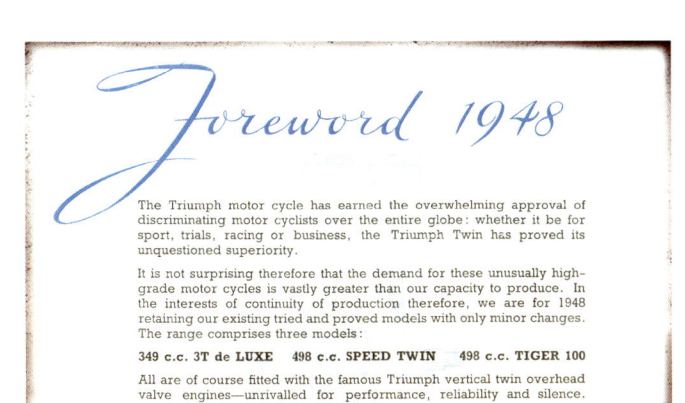

The Triumph motor cycle has earned the overwhelming approval of discriminating motor cyclists over the entire globe: whether it be for sport, trials, racing or business, the Triumph Twin has proved its unquestioned superiority.

It is not surprising therefore that the demand for these unusually high-grade motor cycles is vastly greater than our capacity to produce. In the interests of continuity of production therefore, we are for 1948 retaining our existing tried and proved models with only minor changes. The range comprises three models:

349 c.c. 3T de LUXE   498 c.c. SPEED TWIN   498 c.c. TIGER 100

All are of course fitted with the famous Triumph vertical twin overhead valve engines—unrivalled for performance, reliability and silence.

For perfect steering, roadholding and comfort
TRIUMPH
TELESCOPIC FORKS and SPRING WHEEL

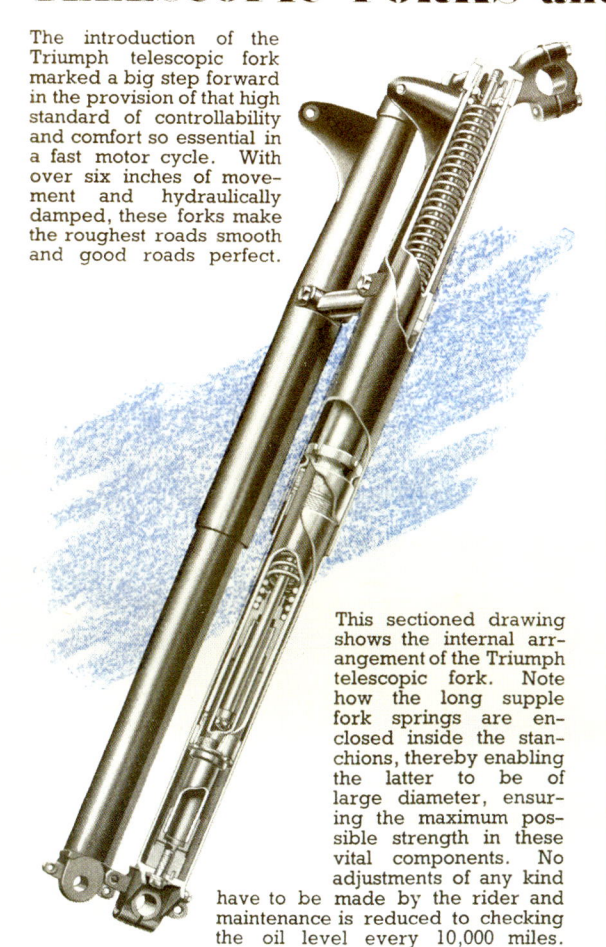

The introduction of the Triumph telescopic fork marked a big step forward in the provision of that high standard of controllability and comfort so essential in a fast motor cycle. With over six inches of movement and hydraulically damped, these forks make the roughest roads smooth and good roads perfect.

From the time that the Triumph Spring Wheel was first demonstrated, it has aroused a tremendous amount of interest among enthusiasts everywhere. Never before has the very complex business of springing the rear end of a motor cycle been accomplished with such a high degree of simplicity and efficiency. It is available as an extra on all three models and when supplies permit will also be available for fitting to earlier types (1937-47).

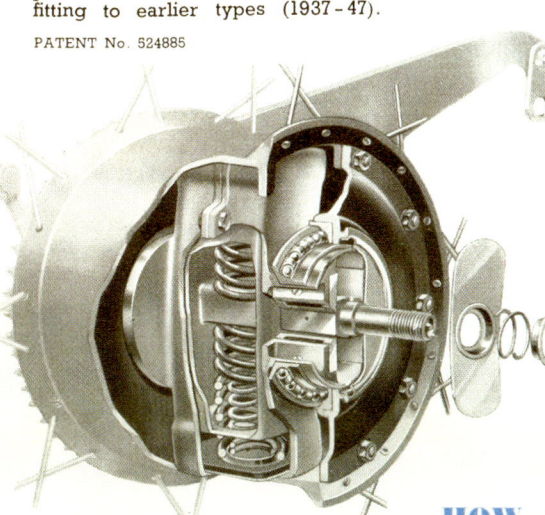

PATENT No. 524885

This sectioned drawing shows the internal arrangement of the Triumph telescopic fork. Note how the long supple fork springs are enclosed inside the stanchions, thereby enabling the latter to be of large diameter, ensuring the maximum possible strength in these vital components. No adjustments of any kind have to be made by the rider and maintenance is reduced to checking the oil level every 10,000 miles.

SIMPLE and PRACTICAL

This remarkable springing system is enclosed in a massive aluminium alloy hub shell to which is attached the powerful eight inch brake. The Spring Wheel is mounted in the frame in exactly the same way as a normal wheel and adds a mere three per cent to the total weight of the machine. Also in the photograph above can be seen the new mudguard design for 500 c.c. models. Simplified staying and elimination of external bolts and rivets results in a greatly improved appearance. It detaches complete from below the saddle.

HOW IT OPERATES

This broken open drawing of the Spring Wheel reveals the essential simplicity and robustness of the design. The spindle remains stationary bolted into the frame as usual while the wheel and hub move on a curved path taken from the centre of the gearbox sprocket which ensures that chain tension remains constant at all times. This movement is controlled by springs, two below the spindle and one above. Lubrication is by a single grease nipple. An important advantage of the Spring Wheel is that the pillion passenger benefits from the springing as well as the driver.

Like the Speed Twin, this T100 from the Sullivan collection came with a four gallon gas tank to ensure a full day's riding without having to stop every half hour for fuel. The one gallon oil tank provided a nice reserve and also helped to keep the engine cool while running hard.

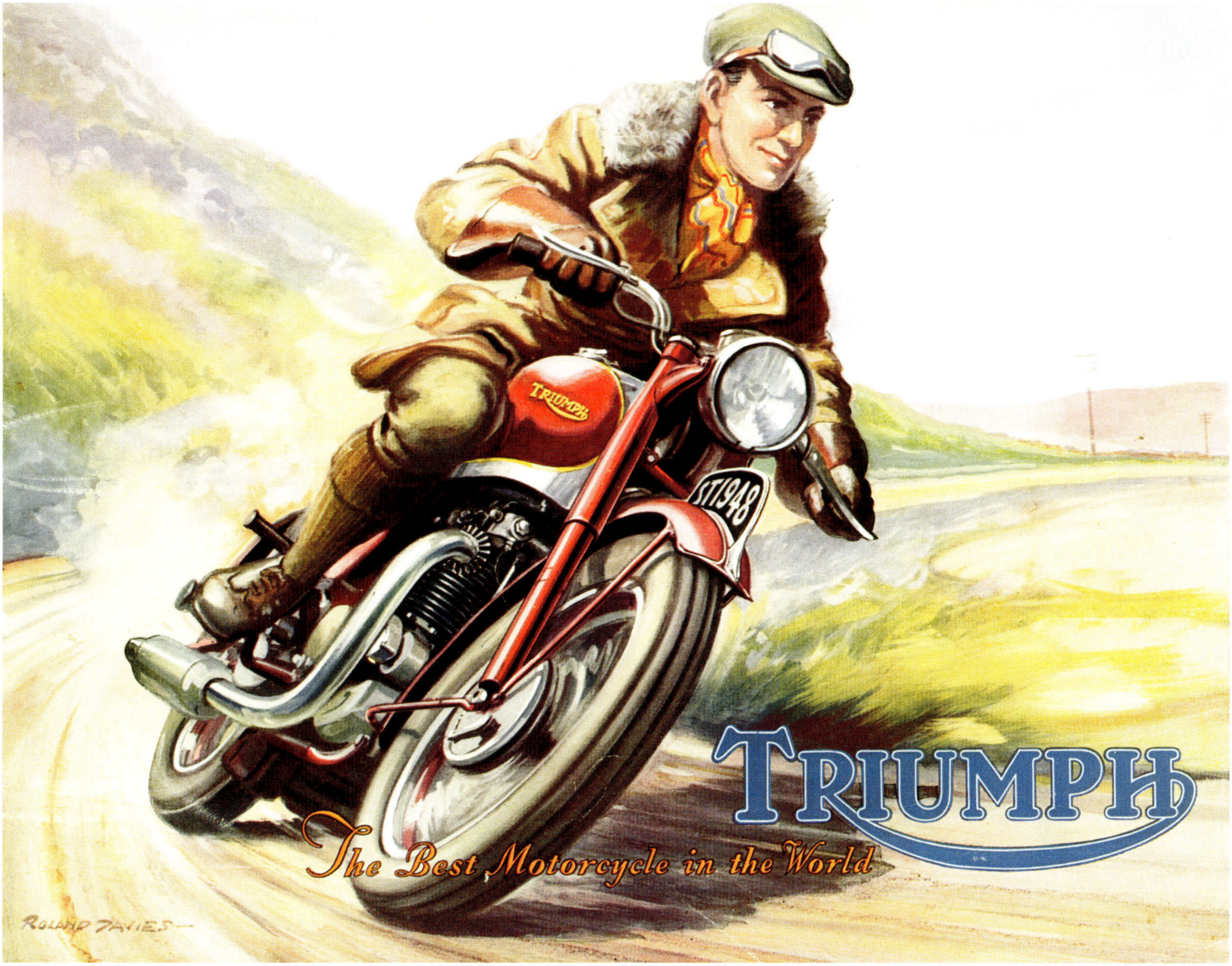
TRIUMPH
ST1948
TRIUMPH
The Best Motorcycle in the World
ROLAND DAVIES

Among the numerous changes that came about at Triumph in the late 1940s and early 1950s, was the redesign of the 500cc cylinder head and cylinder castings. Replacing both the standard cast iron cylinders and the more exotic generator top ends, the new alloy cylinder casting with abundant, closely spaced cooling fins came to be known as the "fine-pitch" engine. In reality it wasn't a new engine, but rather a new top end. These smaller twins would have to wait until 1957 for a complete new powerplant.

Restored by Dick Brown, this TR5 uses the fine-pitch engine mated to a four-speed transmission unchanged from earlier models. The nicely streamlined primary cover hides the standard single row chain used earlier, connected to a clutch assembly with two additional discs. Siamesed exhaust system is nearly identical to that used with earlier generator engines.

Triumph's "new" engine inhaled through an Amal 6 carburetor with typical external float bowl. Aluminum cylinders with steel liners housed silicon-alloy pistons mated to polished Thunderbird connecting rods. To combat the woes of wear, lifters were stellite tipped. The new cylinder head design, and slowly improving fuel of the period, meant the compression ratio could be raised slightly.

No swingarm suspension yet. Only Edward Turner's less-than-satisfactory sprung hub, a troublesome design at best. Most TR5s came with lower gearing and a wide-ratio gear box. Though both the TR5 and T100 came with only one carburetor, Triumph would soon issue a race kit complete with a new cylinder head and two Amal carburetors. Note the parcel grid on the gas tank, a soon-to-be-standard Triumph item.

Even with a new top end, the 500cc Triumph engine could make only so much power. Especially in the American market, cubes were (and remain) king. Increasing the displacement of the Triumph twin to 650cc was a good way to answer the needs of power-hungry Yanks, and keep ahead of other English firms bringing new 500cc twins to market. Seen here are two prime examples of Triumph's early 650cc models. The "blackbird" is one of the rare Thunderbirds painted black at the factory. Mated to the sidecar is a 1954 T110. Both are the work of Dick Brown.

Though sidecars were common in England, where the motorcycle was often the only motorized vehicle in the family, this Swallow side car was seldom mated to Triumphs of the period. On this side of the ocean the extra displacement of the T-Bird and T110 were seen as a way to make motorcycles faster. On the other side, the extra power was often used to pull a sidecar.

The T110 came with more than just additional power. To handle that power Triumph installed a larger 8 inch brake on the front wheel. But perhaps even more important from an enthusiast's point of view, the T110 was the first Triumph to come with true rear suspension. The suspension meant a complete new frame, which in turn gave the designers an opportunity redesign the entire machine. Now the oil tank became a more integral part of the machine, its shape mirrored by the side cover used on the other side. The swingarm frame was in fact a huge leap forward.

Looking at this Thunderbird, it's not too hard to believe that the dealers of the day pleaded with Triumph to paint the bikes in a color other than blue. Whether in black or blue, the big 'Bird was the cruiser of its day. With an under-stressed twin displacing 650cc the bike was easy to ride at slow speeds, yet very fast for the time. The single one-inch Amal carburetor helped ensure the bike would pull well from low RPM. Though the hardtail frame seems an antique today, the bike was 45 pounds lighter than the later bikes with swingarm frames.

The sleek headlight was introduced on the T100 in 1949, and became a standard part of all Triumph's street bikes of the day. Gauges are no longer found in the gas tank, but are instead contained in the new stylized headlight nacelle, along with the light switch. This 1953 Thunderbird represents the end of the hardtail-era, though it does use the Turner-designed sprung hub.

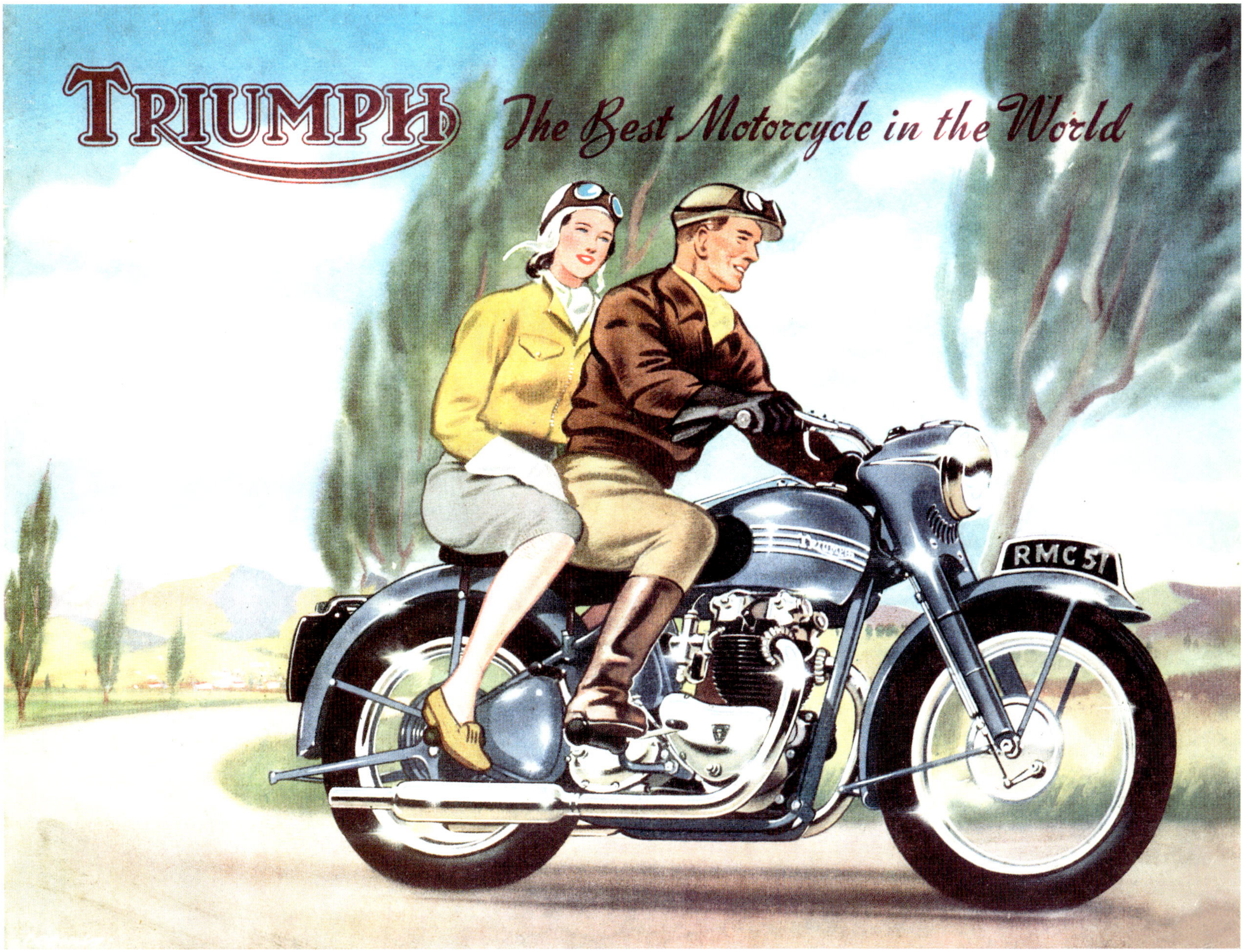
TRIUMPH
The Best Motorcycle in the World
RMC 51

Q&A Randy Baxter

Visitors to the very small town of Marne, Iowa, might be surprised to see a large and prosperous Triumph dealership located along the main road coming into town. Of course most of the first-time visitors to Marne are there because of Baxter Cycle, so the actual number of surprised visitors is few. In fact, Marne is one small Iowa town that actually has a net inflow of individuals everyday, thanks in part to the activities of one hard-working and tenacious individual by the name of Randy Baxter.

Randy started in the business waaay before the current renaissance of interest in Triumph motorcycles began. Not everyone thought starting a Triumph dealership was the smartest thing to do in the mid-to-late 1970s. Luckily for anyone who appreciates old English motorcycles, Randy was listening only to his own inner voice. Apparently that voice said "do it" and Randy did. Today, Baxter Cycle sprawls through the large main building and into two separate and sizable warehouses.

Q&A: Randy Baxter

Randy, how about a little background on you and how you became involved with Triumphs?
My first British bike was a BSA that had been stolen on the East Coast, ended up in the ocean and then made it back to Iowa with a truck driver. That probably wasn't the ideal first British project. Luckily the next Brit-bike was a 1969 Bonneville. That was a way-good bike. After riding that I was hooked. Some of my friends bought Triumphs too and I became the designated mechanic. Before long I was fixing them every night after working my regular job.

When did you decide to open a British store?
Officially, I opened Baxter Cycle in January of 1977, but I kept my day job as a welder and fabricator in a small country shop. In January of 1981 Baxter Cycle became my full time job.

Why open the store in Marne, Iowa, and where the hell is Marne anyway?
It's on I-80, 50 miles east of Omaha. I bought the place because there was a shop with a house next door and it was something I could afford.

Did you ever think the business would get as big as it has? And do you have plans to expand the store?
No, I never dreamed it would do anything like this. A lot of people in the motorcycle business said I was an idiot to

Q&A Randy Baxter

have anything to do with British bikes. Funny, a lot of those people aren't in the business anymore and I still am. In terms of an expansion, it's in the works. We're going to expand and modernize the showroom. We're dealers for Triumph, Royal Enfield and soon-to-be Norton, so we need more space.

Where do you get your bikes, and where do you get your parts?

Bikes come from all over the states. Auctions, swap meets, trade ins. People call us with motorcycles they want to sell, because they know we're here. The parts come from all over the world. We source them everywhere from Europe to the machine shop three blocks away. They come from China, New Zealand, Australia and all over the US. We have some parts made for us, like the rod bearings that are manufactured in Atlantic, Iowa, just east of here.

You seem to have a million parts on the shelf, are they inventoried on a computer or is it something you have in your head?

The parts are on the shelf in numerical order, so it's possible to check and see if we have something or not.

How many finished classic bikes do you keep in stock, and what are the brands you handle?

At any given time we have between 75 and 85 compete bikes in the showroom. We handle most of the English brands, including Triumph, Norton, BSA, Matchless, Royal Enfield, AJS Velocette and Vincent.

What do you see happening to the value of nice original or restored Triumphs?

The prices keep going up. Ten years ago if you told me a sixties Bonneville would sell for eighteen thousand dollars I never would have believed you. It's not all baby boomers either. Mostly they fuel it, but some younger guys are getting involved too.

Military bikes were an integral part of Triumph production, even after the end of WWII. Because the tooling for their new military bike, a 350cc ohv twin, was destroyed in the air raid of November 1940, Triumph relied on the 3HW single cylinder bike throughout most of the war.

When the time came to design another military bike, the engineers at Meriden came up with the TRW. Rather than a clean sheet approach, the design staff chose to convert a typical Triumph twin from overhead valves to the simpler flathead design.

Production of the new TRW bike was delayed by the war it was designed to serve, which meant the bikes didn't really go into production until the very end of the WW II. This particular example from Baxter Cycle was originally sold to the Canadian military and carries a 1953 production date. Most of the parts are converted from civilian bikes, right down to the oil tank and tool box. Note the post-war hydraulic fork, headlight nacelle and gas tank.

In an attempt to cover the entire motorcycle market, Edward Turner directed the design staff to create a small, attractive commuter bike. The T15 Terrier mimics the looks and lines of the bigger bike, and actually looks much bigger than it's 150cc displacement would suggest. The engineers and designers worked hard to integrate the bike with the rest of the Triumph line, note the headlight nacelle and the gas tank complete with knee pads.

Part of what makes the Terrier look so convincingly like a bigger bike is the large diameter 19 inch wheels with 2.75X19 inch tires at either end. Despite the big-bike looks, the Terrier suffered from reliability issues and never sold in huge volume. The best thing about the Terrier might be its offspring, the single cylinder Cub, one of the more successful bikes ever produced by Triumph.

Owned by Ray and Le Ellen Corlew, this low-serial number Terrier (and thus likely a 1953 model) uses a plunger system in lieu of what most of us would call real rear suspension. If the suspension was old school, the engine was very modern for the day. The ohv design uses a cylinder and head die cast in aluminum. All the castings, from the head to the cases, share that nice form-follows-function beauty seen in so many Triumph parts. Alternator electrics dictated the rectifier seen under the seat.

TRIUMPH TIGER CUB
Patent Nos. 475860, 474963, 482024

To the sportsman the "Tiger Cub" makes an instant appeal. It is a sporting lightweight designed for brisk performance, with safe easy handling, good brakes and comfortable suspension front and rear. Beautifully finished in shell-blue sheen and glossy black like the other Triumph "Tiger" models.

SPECIFICATION

BRAKES. Large diameter cast iron drums. Exceptionally powerful yet smooth and safe in use.

WHEELS. Special Triumph design with dull-plated spokes and chromium-plated rims. Dunlop tyres. Efficient mudguards front and rear.

ELECTRICAL EQUIPMENT. The well-proved A.C. Lighting-Ignition system with crankshaft mounted alternator and emergency start circuit. Large diameter headlamp and powerful rear lamp with combined reflector.

TOOLBOX. All-steel with secure fastener. Complete kit of good quality tools.

NACELLE. Triumph patented design, integral with top of the forks. This neat streamlined shell encloses the headlamp, instruments and switchgear. Also includes a gear position indicator.

SPEEDOMETER. Smiths speedometer mounted in nacelle, internally illuminated.

OTHER DETAILS. Finish: T.15 Amaranth Red; T.20 Shell-Blue sheen and Black. Quick action twistgrip. Rubber knee grips. Upswept exhaust pipe optional on T.20.

PAGE THREE

P. H. Alves, famous Triumph rider, on his "Tiger Cub". ("Motor Cycling" photo.)

Perhaps one of the best looking of the pre-unit Triumphs, this 1955 T110 is the work of Baxter Cycle, who rescued and restored the bike after a 25 year rest in an old shed. The T110 remained Triumph's hot cruiser right up to the introduction of the Bonneville. Though still sporting only one carburetor the bike was a rocket in its day. Twin carburetors for the 650cc line would have to wait for a new cylinder head - a twin carb race kit was made available in 1958 and came factory-installed on the Bonneville in 1959.

With the exception of the aftermarket risers from Flanders, the bike seen here is 100% correct. The long, one-piece seat was introduced when the line went to the new frame with true rear suspension. The T110 was the first Triumph to come with an eight inch front brake, the effectiveness of the brake meant it soon was added to other members of the Triumph family as well. Equipped with a generator, this bike uses the early-style primary cover. Note too the magneto just behind the cylinders, points-ignition was still years off.

Though the tall risers are non-stock, they are very much a common addition from the period, and were on the bike when it was found in the shed. 1955 was the last year for the cast iron cylinder head on the 650 twins, though the 650cc bikes would continue to use a separate engine and transmission for some years to come.

Not the most attractive color ever used on a Triumph, this Jim-Hess restored TR6 dates from 1956, the first year Triumph put a 650 engine in a Trophy frame and offered it for sale to the public. Note the aluminum head and the siamesed exhaust system.

It's hard to succeed in business without taking risk. The original Speed Twin was just such a risk, one that paid off very well for Triumph. In the mid-1950s they took another, by redesigning the Speed Twin, the T-bird and Tiger 110. The redesign enclosed the back of the bike in what looked to Americans like an inverted bathtub. Taking the theme forward, the front fender became a skirted affair with a flair at the bottom edge.

Mountains of bathtub enclosures piled up behind most US dealers, because in order to sell the bikes, they had to be converted to non-bathtub status with conventional front and rear fenders. Stock bikes, like the Speed Twin shown here are thus rare as rare can be.

This particular bike, the work of Baxter Cycle, dates from 1959 and uses the nearly-new unitized 500cc engine and four-speed transmission. To quote Randy Baxter, "The idea was to build a commuter bike for drizzly climates. But over here people couldn't get that sheet metal off fast enough."

Unlike modern bikes which require very little regular service, motorcycles of the day were mechanical affairs that required a fair amount of maintenance. Today, when there's a flat tire or breakdown, the first tool we reach for is the cell phone. Triumph riders of the 1950s had no such luxury and were sometimes forced to effect roadside repairs. For such a rider the tool kit was as essential as gasoline. The kit for this bike, made up of Triumph-specific wrenches, is tucked away neatly under the one piece seat.

Parked under the wing of a Pitts S1, these two Triumph cubs show just how much Triumphs, and the world, changed between 1959 and 1969. While the earlier bike is pure Triumph, by 1969 BSA played a big part in the design and manufacture of the bikes.

While the older cub uses the engine first seen in the Terrier, the later Trophy 250 borrows heavily from the BSA line to offer a bike with the looks and stature of the bigger twin cylinder bikes. Even the twin-leading-shoe front brake was used on the smaller machines.

650cc & 2 Carbs

Triumph's hottest new model, the one we all identify with Triumph and the one many of us remember from "the day," wasn't truly born in 1959. The Bonneville's genes go back at least to the early 1950s and the introduction of the 650cc Thunderbird and the T110. Considering that the dual carb head was a huge part of what made the Bonneville special, the big Bonnie owes part of its success to the first hot rod 500cc models. As early as 1953 Triumph offered a T100C with a twin-carb cylinder head and high-lift cams.

The evolution of the Bonneville took another leap forward in 1956 when Triumph introduced the new aluminum Delta head for the 650 twins. This new head allowed the engineers to run more compression and also eliminate the external oil drains. With the new improved 650 engine and a tried and true TR5 frame, Triumph created another home run, the TR6 - the first 650cc Trophy. And though the new bike was sometimes referred to as a Trophy-Bird, the engine used in the TR6 was the high output T110 model, not the Thunderbird mill.

The new TR6 came to the showrooms with a single stand-alone chrome headlight and simple non-skirted fenders. The bare bones look appealed to most American buyers. With its simple, clean Yankee design, the TR6 made the first year Bonnie look rather dated and dowdy. By contrast, the headlight nacelle and skirted fenders used on the first Bonneville made it look more like an old man's cruiser than the object of a young man's lust. But all that was about to change.

Looking very mechanical, this second-year TR6 from 1957 uses the T110 engine in the Trophy frame, without any of the sheet metal encumbrances seen on many of the other Triumphs from the same period. The left side shows the primary cover with the simple dimple for the crank end, a design that would change with the introduction of the alternator in 1960. Note the attractive off-road two-into-one exhaust.

As the TR6 started life as a Trophy, or competition bike, the model carried minimal equipment and sheet metal. Just the bare necessities to go down the road or around the race track. Not only was the bike a winner at all kinds of off-road events, it also set the style for the Triumph's best known bike, the Bonneville.

The bike we've all heard so much about looks somehow less-than-sexy in it's first-year livery. The headlight nacelle meant there was no easy way to mount a tachometer, and though the bathtubs were long gone, the big heavy fenders just weren't what you expected to see on Triumph's answer to other hot bikes of the day, including Harley-Davidson. Anything the engineers got wrong in the styling department they more than made up for with the engine. The first Bonneville came to the market with nearly 50 horses, thanks to 8.5 to 1 compression, two carbs and sport camshafts.

Edward Turner was a man blessed with many abilities, marketing was just one of them. The Bonneville name, based on Triumph's records on the salt flats, was nothing short of genius - at once both a very American name with all kinds of hot rod associations, and a reminder that Triumph held the title for the world's fastest motorcycle. Though the twin Amal carbs came with mini velocity stacks, they didn't come with aircleaners.

Nothing epitomizes the change between the 1959 and the 1960 Bonneville better than the headlights and gauges. Though the nacelle on the 1959 model declares the Triumph a "world motorcycle speed record holder" the 1960 model dispenses with the headlight nacelle altogether. In its place the later bike offers a big, chrome and very American, headlight housing complete with small built in ammeter. Equally important is that second gauge mounted alongside the speedometer, the tachometer. Now we have the makings for a proper hot rod.

Especially with the low bars, this 1960 Bonneville owned by Bobby Sullivan looks ready to race. In addition to the obvious lack of ugly sheet metal, the 1960 model differs from its earlier kin by the addition of a crank-mounted alternator and a new bulge in the primary cover. The crank itself is a one-piece affair (actually introduced in 1959) designed to withstand the rigors of the twin-carb engine. Because the 1959 model didn't sell as well as anticipated, there are some "1960" Bonnevilles that are actually re-titled 1959 bikes, complete with the extra sheet metal.

In 1959 the TR6 looked better to most American riders than the hotter Bonneville, especially with the fat 4.00X18 inch tire in the rear mounted to a chrome rim (up front a 3.25X19 led the way). The TR6 looked so much better than the Bonneville to some riders that they bought the single carb bike and converted it to twin-carb status with a kit from Triumph, first made available in 1958. This pre-alternator bike uses the generator at the front of the engine, the two-into-one exhaust and the tachometer driven off the left side of the exhaust cam. Jeff Hackett photo

This 1960 TR6 from Baxter Cycle shows the other face of a TR6. With standard ribbed tires and more typically Triumph exhaust and mufflers, the TR6 made a great street bike, one that was very nearly as fast as the Bonneville. 1960 saw the beginning of the new duplex frame with twin downtubes. Not Triumph's best design, these frames were known for vibration, cracks near the steering head and cracks to the gas tanks as well.

Despite a few issues, like the new frame, Triumphs of this era were seen as good bikes of high quality and Triumph sold as many bikes as they could comfortably produce. Millions of Baby Boomers were just reaching driving age and most of the male half of that boom longed for motorcycles. Luckily for Triumph, their 650cc machines occupied the top spot on many of those young mens' wish list.

Have you ever seen a better looking motor? (maybe a nicely polished Shovelhead). With the rare high pipes and polished cases, this Bonneville, owned by Bobby Sullivan, exemplifies all the reasons for Triumph's popularity. The bikes were fast, fun to ride and very good looking. As many have noted, each part on a Triumph looks like it belongs there with all the others. And yet, each one can stand alone as a great design with a pleasing shape.

The evolution of the big 650cc Triumph twins moved at a sedate pace in the early 1960s. Other than the duplex frame and alternator, the bikes changed very little from year to year. This particular T120 is a competition model, which explains the high pipes, and the hard-to-see quick detach headlight. In many way these early competition bikes were the precursors to the later and legendary TT models.

FOR THE PACKAGE OF POWER

GO Triumph

POWER — is the magic word for '62! In the new Triumph line, the emphasis has been put *inside*, where you can *feel* it, where it counts! That's why riders who want ACTION will be choosing the '62 Triumphs — the High Performance OHV line with the "Package of Power."

*In speed there can be no argument. Triumph holds the A.M.A. approved WORLD'S ABSOLUTE SPEED RECORD, Bonneville, Utah, 214.47 mph (with streamlined shell) plus the A.M.A. Class AA record of 159.54 mph (not streamlined).

TR6S/R TROPHY ROAD SPORTS
40 cu. in (650 c.c.)

A single-carburetor, high performance road model. Features Trophy front forks and detachable headlamp. Also available in Competition form. New *Flamboyant Ruby Red* and *Silver Sheen* with *Black* frame.

6T THUNDERBIRD — ROAD TOURER
40 cu. in. (650 c.c.)

For the road rider who wants top power combined with extraordinary reliability and economy. Offers maximum comfort, safety and fine handling. *Kingfisher Blue* and *Silver Sheen* with *Black* frame.

A MAN'S motorcycle for a MAN'S Sport . . .

Motorcycling *is* a man's sport. It takes stamina, endurance, and skill. It's good, clean fun, and takes you out-of-doors, helps develop qualities of independence, self-reliance and quick judgment. The motorcycle has to be able to take it — and give it. You need a rugged machine that can take it, but also one that can get up and *go* when you "pour it on." For motorcycling fun, nothing beats a Triumph!

Triumph offers the U. S. rider what he wants, at a price he can afford to pay.

T120/R BONNEVILLE ROAD SPORTS
40 cu. in. (650 c.c.)

For the expert rider. Famed OHV Vertical Twin with twin carburetors, known as "The Fastest Standard Motorcycle made in the world today."* Many mechanical refinements. Also available in Competition form. New *Flamboyant Flame* and *Silver Sheen* with *Black* frame.

T100S/C TRIUMPH ENDURO TROPHY
(Supersedes TR5A/C) 30.5 cu. in. (500 c.c.)

New mechanical improvements give this famous Trophy even greater performance and handling ease. For competition riders. *Kingfisher Blue* and *Silver Sheen* with *Black* frame.

T100S/R TIGER ROAD SPORTS
(Supersedes TR5A/R) 30.5 cu. in. (500 c.c.)

This top favorite, introduced last year, promises even *more* for '62. High performance, unit construction over-square engine. *Kingfisher Blue* and *Silver Sheen* with *Black* frame.

By 1962 the TR6 was fast becoming a single carburetor Bonneville. Engine changes were few for that year, though the crankshaft balance factor was changed which required a new counterweight. This 1962 TR6SS was discovered by Jim Hess with only 11,000 original miles on the clock. The low mileage bike was then taken to Kenny Dreer for a complete restoration.

Though overshadowed in the US market by the TR6 and Bonnie, Triumph continued to manufacture the Thunderbird and T110 models. And as this model shows, the bathtub-style enclosures didn't go away either. The year 1963 was huge for Triumph, the list of changes and improvements was long and impressive. In place of the duplex frame, Triumph brought out a single downtube frame. Instead of the separate engine and transmission, they brought out the unitized engine and transmission connected by a new, stronger duplex primary chain.

What could be better, a hot date on a hot motorcycle. Life really was good! Up front a 1964 Bonneville and in back a white 1963 model. Both use the single downtube frame with improved rubber mounting for the gas tank and an oil tank with hidden filler cap. The biggest news for 1963 though was the unit construction engine and transmission with the new timing-side and primary covers. Taken together, the multiple changes introduced in 1963 made the 650cc Triumph a much more sophisticated motorcycle. Both machines from the Sullivan collection.

It wasn't until the middle of the 1963 model year that the Bonneville received air cleaners! The new bike retained all the signature Triumph design elements: chrome rims, large chrome headlight shell, twin Smiths gauges, graceful tank with package grid, and overall a certain sense of style and balance. The new engine was really a blend of old and new. The dimensions and location of critical components like camshafts were essentially unchanged. Up top, the cylinder head was new, as were the rocker boxes with their stylish horizontal fins. Ignition was now triggered by points wired to coils tucked up under the tank.

Though the 1963 Bonnie came to town in boring white, the sibling TR6 from the same year rolled out of the crate in bright regal purple sprayed over the base silver for even more brilliance. With all the changes Triumph made in 1963, they chose not to change the tank badge. Like the Bonnie, the 1963 TR6 uses an 8.5 to 1 compression ratio and new 4819 intake and 4855 exhaust camshafts. The dual ignition points were driven off the right side of the exhaust cam, which meant the tachometer drive, formerly driven off the right side of the exhaust cam, had to be moved to the engine's left side. Part of the Sullivan collection, this bike wears perfect paint by Perewitz.

The missing generator and magneto, combined with the unit design, makes for a much cleaner and modern looking engine, which in turn helped modernize the whole machine. The sleek twin seat, simple taillight, sensuous tank shape and abundant chrome helped keep Triumph at the top of the sales charts throughout the 1960s.

By 1964 the Bonneville came standard with air cleaners. The unit engine and transmission meant a much tidier visual package especially on the right side. No longer is there a big gap between the transmission and the engine cover. Gone too is the oil line that ran along that gap on the pre-unit bikes. After all the changes made for the 1963 bikes, the pace of change slowed during 1964 and 1965. Perhaps the biggest change for 1964 was the introduction of a new improved fork assembly with external springs.

Restored by Baxter Cycle, this 1965 Bonneville shows off the great lines that made a Triumph one of the world's most most sought-after bikes, then and now. Little things, like the shape of the side cover, the chrome front hub and the cast-in Triumph logo in the primary, make a big difference when taken collectively. At the time, the only thing as cool as a 1965 Bonnie would be a Harley-Davidson Sportster. The two bikes were constant rivals when the light turned green and started many an argument in biker bars of the period.

One of the better colors from the period, the pacific blue paint is laid down on top of the silver, which makes for a nice color combination while helping to brighten the blue with a lit-from-behind candy effect. Larger rubber gaiters seen on the forks, as compared to earlier bikes, are part of the move to external springs.

Other than the fork, most of the 1965 bike is the same as the 1964 model Bonneville. The gas tank and egg-crate tank badge are straight from 1964, though the tank badge would be replaced one year later.

Owned by Mark Jensen, this very rare C-model TR6 from 1964 is one of less than 200 manufactured at the Meriden plant. Note the perfect chrome on the very hard to find signature high pipes. This TR6 came with the same new "external spring" fork seen on Bonnevilles from the period. The front drum is the same one used one year before.

Because the bike is a C-model, it came with a number of unique features like the missing tachometer, high pipes and the rigid-mount handle bars. Otherwise the bike is a standard-issue TR6 with a 650cc engine equipped with one Amal Monobloc carburetor.

Restored by Gary Chitwood for the Sullivan collection, this 1965 TR6 is ready for the road, complete with speedometer and tack, and low roadster pipes and mufflers. Most of the changes for 1965 were very minor, like the mounting of the horn and the new taillight from Lucas. The rear hub is new as well, though the rims and wheel diameters remained the same as those used the year before.

For the second year of production the external-spring fork came with longer compression springs and shorter top-out springs, for one extra inch of travel. Otherwise the 1965 TR6 bikes were very similar to those produced one year earlier. The tachometer is driven straight off the left side of the exhaust cam, which means there's no good way to route the cable without bending it into a sharp turn. Instead of silver, the basecoat color is Alaskan white, topcoated with burnished gold for the upper half of the tank and in a stripe that runs down the middle of the fenders.

By 1966 things were happening again at Triumph. Some of the changes seen on this Baxter-built TR6 include the new alloy taillight housing - though the lens is the same one used one year earlier. The other big visual change is the new eyebrow-style tank badge which replaced the long-running egg-crate design used earlier. Color is pacific blue over Alaskan white. Seen in the background is a 1967 TR6.

Some of the biggest changes introduced in 1966 are hard to find or see. The lighter crankshaft assembly, supported by a new primary-side roller bearing, was designed to let the engine rev more quickly. Both the TR6 and Bonneville used the 4819 intake and 4855 exhaust camshafts operating R-series lifters. TR6 bikes used a single, 1-1/8 inch Amal Monobloc carburetor. Equally hard to see is the electrical system and battery, which were finally converted to twelve volts.

For 1966 Triumph unleashed a whole series of changes and improvements. Things like the new eyebrow tank badge, wider front drum for more brake area and slimline gas tank. The gray grips were used only at the start of the model year, so Denny Narland installed them on this low-serial-number bike. Carburetors remain the tried and true, and much maligned, Amal Monoblocs. Bikes delivered early in the year came with 1-1/8 inch carbs, though later bikes used Amals with a 1-3/16 inch bore. Safety wire was no longer used on the three carburetor screws.

To help the Bonneville live up to its hot rod reputation, the 1966 bikes used R-series lifters and more aggressive E4819 intake and E4855 exhaust camshafts, along with the bigger carburetors already noted, and a lighter crankshaft assembly. Static compression was raised as well, from 8.5 to 1, to 9 to 1.

Unlike the big twins which went to unit construction in 1963, the smaller twins became unitized in 1957. Introduced as a 350cc, it wasn't long before the engine was bored out from 58.25mm to 69mm, which when combined with a stroke of 65.5mm made for an over-square 500cc engine. This new 5Ta became one of the winningest of all the Triumph powerplants. With single Amal carb, 9 to 1 compression and two-into-one exhaust the 1966 T100 SC was popular on the enduro circuit. In fact, the T100 SC won the National championship in 1962, 1963 and 1964.

Starting in 1964, the T100 SC used aluminum fenders in place of painted mudguards. The gas tank is new for 1966, as is the fully rubber mounted oil tank. The front fork is the same external spring unit used on TR6 and Bonneville models of the same period. This off-road Triumph came with a speedometer and no tachometer, a small-diameter headlight and the siamesed left-side exhaust.

Throughout the 1960s, the Bonnevilles and TR6 bikes became more and more refined. As part of that eternal quest for more power, the 1967 Bonnies used Amal carbs measuring a full 1-3/16 inches. Though the bike shown uses Monoblocs, later bikes from 1967 used Concentrics. Combined with the hot cams and 9 to 1 compression introduced one year earlier, this 650cc twin put out a claimed 52 horse power.

Left side view shows the tachometer cable, now driven through a right angle adapter for much neater cable routing, and the headlight switch missing from the side cover and moved to the headlight housing instead. For 1967 the fenders are stainless steel, the paint color is aubergine over Alaskan white (earlier bikes used gold in place of white). The combination of deep red and white paint, along with the typical and abundant chrome and polished aluminum is hard to beat.

The TR6 bikes came in a variety of "models," including this 1967 TR6C. Designed for Western off road events, the bike uses a 3.50 inch front tire and 4.00 inch Dunlop in the rear. Note the skid plate under the engine, and the rear-wheel-drive for the speedometer (introduced in 1966). Vertical rib cast into the cylinder head was designed to prevent the fins from playing a harmonic song at certain rpms.

Though meant as a special competition model, the C came with the same single-carb 650cc engine as all the other TR6 bikes. According to racing heroes like Bud Ekins, torque and durability were the keys to winning western race events, not brute horsepower. The C model TR6 uses the same smaller tank seen on Bonnevilles of the period, minus the parcel grid eliminated one year earlier.

The biggest change in the 1968 bikes is the new twin-leading-shoe front brakes. By using one cam per brake shoe, each positioned so the expanding shoes would self-energize with the application, the Triumph engineers designed a drum brake with the potential to be almost as good as a disc brake. The only trouble of course is the long cable and poor routing used in 1968. Riders complained of a spongy feeling lever and less than terrific performance, a problem that Triumph resolved one year later with a shorter cable and improved routing.

From this perspective its easy to see why Triumph designated these two-and-a-half gallon gas tanks as "slim line." With the limited capacity they might have been impractical, but they sure were sexy. Thin, glue-on knee pads and die cast tank badges exaggerate the tapered shape of the tank. By 1968 the ignition switch migrated to the left side of the fork. From this angle it's also easy to see why the twin-carb heads were known as a splayed-port design.

This TR6C, from 1968, is owned and maintained by Dennis Beckman. Like the Bonneville, this TR6 uses an improved fork with damping on both compression and rebound. By 1968 the TR6 used an engine with only one carb, but was otherwise identical to that found in the Bonnevilles. Both models used 1-19/32 inch intakes and 1-7/16 inch exhaust valves operated by 3134 camshafts for both the intake and exhaust.

Attractive and hard to find left side pipes were used on only a few of the TR6 models. Note the small-diameter headlight with chrome housing, and the streamlined zener diode, designed to bleed off excess alternator output, mounted between the fork legs.

The 500cc bikes are often overshadowed by the bigger, sexier 650cc models, especially in the US. That doesn't mean the "little" Triumphs weren't fast. This 1968 Daytona, restored by Dick Brown, comes by its model name honestly, as twin-carb 500cc Triumphs won Daytona in both 1962 and 1966. Not to mention innumerable wins against the bigger Harleys at TT and flat track races. You could even say the 500cc bikes served as mentor for the 650cc machines: they were first to use unit construction, points ignition, and a "delta" twin-carb head.

Very similar to the TR6C seen a few pages back, this TR6 from 1968 is part of the Bobby Sullivan collection. In place of the high pipes, this bike uses standard roadster pipes meant for street use. And rather than bolt on a small headlight, this bike uses the same headlight housing found on Bonnevilles of the period. By 1968 the switches were gone from the side cover, moved to the headlight and fork-ear instead. Amber and red side reflectors are part of the Federal mandates for bikes manufactured in 1968. Note the subtle height difference in the seat.

The 650cc machines continued to evolve right up to the end. This 1969 TR6 from Denny Narland shows the improved routing of the front brake cable, the one thing that twin-leading-shoe design needed to actually work as intended. By 1969, both models of the 650cc line used the same engine, the only difference being the number of carburetors. For 1969, the engines came with a heavier flywheel but were otherwise much the same as those produced one year earlier. Note the exhaust cross-over tube, designed to improve mid-range power.

Though we think of it as a TR6, Triumph called this a Tiger 650. New seat rail is essentially a piece of visual candy. Anyone who tried to use the rail to lift the bike onto the centerstand only did it once – then they looked for a way to repair the seat trim and pan. By 1969 the sale of Triumph to BSA, an event that occurred almost 20 years earlier, was beginning to show its ugly side. During this time more and more parts were shared between the two marques, and many parts were converted to "unified" threads and dimensions.

Variations on a theme. This Denny Narland owned and detailed TR6 is a C model from 1969. He calls this an "unmolested bike," one of the few that needed only a re-painted gas tank and new aftermarket exhaust to look as good as it does. Like most C bikes, this one uses a small headlight, no tachometer and high pipes running along the left side. Though many TR6s use the bigger gas tank, this model sips fuel from the skinny slim-line "Bonneville" tank.

The new heat shields weren't real popular with the Triumph faithful, many of whom called these the "barbecue grille shields." High pipes use a cross-over pipe like that on the roadster exhaust of the period, but located just ahead of the mufflers. While the road ready TR6s and Bonnies used painted fenders, these are stamped from stainless steel. With the simplified gauges and competition orientation, this TR6C is much more like the early Trophy bikes than the standard issue TR6R.

A pair of Bonnevilles, a 1968 owned by Denny Dingman in the foreground, and a 1969 owned by Wendy Parson bringing up the rear. Differences include the abbreviated eyebrow tank badge which showed up in 1969. The scalloped paint job seen on Wendy's bike is one of only three paint jobs introduced at the end of the wild and crazy 1960s. Bar-end mirror and non-crossover exhaust pipes are aftermarket additions to the 1968 model.

Another 1969 Bonneville, this one owned and restored by Denny Narland. Note the "hockey-puck" horns (some 1969 bikes used a domed horn) and the Amal Concentric carburetors, added in place of the Monoblocs mid-way through the 1967 year to both the TR6 and Bonneville. The Concentrics used both a "tickler" the small brass pushrod seen on the side of the carb, and a choke.

A light, simple motorcycle, sign of a simpler time. No electric starting, and no need for the big heavy battery, no turn signals and no hydraulic disc brakes. In fact, the drum brakes used on the 1969 and 1970 bikes were pretty good – almost as good as a disc. But the Japanese were knocking on the door with better, faster motorcycles. Bikes with better brakes, push button starting and far fewer oil leaks. At Triumph, there didn't seem to be anybody home.

Two 1970 models from the Sullivan collection, a TR6R in front and Bonneville behind. Both bikes use the new primary breather, and both contain a wealth of unified nuts and bolts in place of the British Standard Coarse (Whitworth) and British Standard Fine Cycle Engineers Institute fasteners. While the TR6 used the three-and-a-half gallon touring tank, the Bonnie came with the slim, definitely-not-for-touring two-and-a-half gallon tank. Some riders preferred the TR6 of the day as it was considered simpler, less stressed and a little more trouble free.

Do it in da dirt. This TR6C from 1970, restored by Baxter Cycle, was a good street scrambler in its day, happy on almost any terrain. Like earlier C bikes it wears a small 5-3/4 inch headlight and the high, left side pipes with the unpopular heat shields. By 1970 the chrome seat rail was moved to the rear sub frame, the location that should have been used from the beginning.

Though the TR6 bikes claimed 45 horsepower, and the Bonnevilles a full 52, neither bike could claim to be king of the hill. The fact that the Speed Twin engine, introduced in 1937, was still in existence and selling in goodly numbers thirty three years later is an amazing fact. But to bring the bikes into the new decade would take a thorough redesign. Unfortunately, the new BSA management team was busy assembling a new R&D center, the thrust of which would be the redesign of all the things that didn't need fixing. Rather than work on vibration, oil leaks or reducing warranty claims, the BSA management chose to work on the frame and the styling of the classic Bonneville and TR6. In their spare time they designed a three-wheeler that was never produced.

Two Magic Letters

Before there were TT bikes there were competition models of the Bonneville. In 1960 they called the two models A and B, with the B equipped as a scrambler. In 1961 the model designations were changed to R and C, a more logical way to identify the Road and Competition bikes.

In 1963 Triumph offered the first true TT bike. Though built in England, it was clearly meant for the US market. The new TT machines took a no-holds-barred attitude toward competition. While the 500cc "Daytona" bikes were consistently placing in the winner's circle at road-race events, the TT bike (and close kin TR6SC) were often the first across the line at flat track and TT events. The bikes were so good that riders often rode them in as-delivered condition, and won.

The TT model was the brain child of Triumph's Western US distributor, Johnson Motors (aka JoMo). The TT bikes were both less and more than a standard Bonneville. Less in the sense that there was no speedo or lights. More when it came to the engine. Which enjoyed more compression, more carburetion, and more duration and lift in the camshafts.

As the Bonnevilles evolved and improved during the period 1963 to 1967, so too did the TT bikes. Thus the TT bikes acquired a better fork, a wider rear swingarm and reinforced swingarm pivot point when the road-ready bikes did.

Why the winning and very popular design came to a stop in 1967 no one seems to know. Which makes the remaining TT bikes (especially the real ones) more valuable still.

"Power, power, power," the mantra of bad Sunday-morning radio advertising, might be the catch phrase for the entire TT line. When Triumph and JoMo got together to develop a hot machine, they didn't stop halfway. Anyone racing a TT bike, like the Chitwood-restored and Sullivan-owned 1963 model shown here, needed a strong right leg, as the compression was bumped all the way to 12 to 1. And not a compression release in sight.

It takes more than compression to make power. While the standard Bonnies used a 1-1/16 inch Amal Monobloc, the TT bikes came to the track with 1-3/16 inch Amals. Considering the fact that these bikes ran in off-road events, it's fortunate that Triumph introduced air cleaners for the twin carb bikes starting in 1964. To open and close the Bonneville-spec valves Triumph specified the 3134 camshaft for both intake and exhaust. What we think of as TT pipes didn't actually appear until 1965. Fenders used in 1964 are light weight aluminum, not stainless.

Bikes built before the introduction of the short TT pipes like this 1964 model from the Sullivan collection used the high pipes seen on the competition Bonnevilles from the early 1960s, but with a straight pipe in place of the small mufflers. Rims are the same as those used on the street bikes, but with a wider 3.50 inch Dunlop in front and a 4.00 inch in the rear. Small rubber plugs replace the switches in the left side cover, and fill the parcel grid holes in the tank. Gearing for most of the TT bikes was lowered from the 4.84 to 1 used by Bonnies to 5.41 to 1.

Painted in Pacific blue over silver, (Bonneville colors) these Dick-Brown restored bikes from 1965 are among the best looking of the TT bikes. The fenders are aluminum, though some TT bikes from the same year came to the dealers with painted fenders. To ensure the TT bikes had enough spark without the need for a battery, Triumph used their Energy-Transfer ignition, powered by special alternator windings and triggered by a standard set of points. The unique alternator used permanent magnets and required no battery, which kept the bikes simple and light.

Climbing a hill or running at the head of a pack of flat trackers, there weren't many races a good TT bike couldn't handle – and win. Like the year before, the 1966 TT bikes came with either aluminum or painted fenders, depending on whether they were delivered through the East or West Coast distributors. And like all 1966 bikes, this one wears the new tank badge affixed to the new slim line tank used on the Bonneville line. Paint is also the same as that used on the Bonnevilles of the same year.

Owned by Denny Narland, this last-year TT bike was discovered in as-is condition. What Denny calls, "one of the few survivors." In 1967 the frame for the Bonneville, TR6 and TT came with a 65 degree steering head angle. Forks used on the TT bikes were built to Bonneville specifications, though the springs were stiffer to better handle the rigors of racing. Though still equipped with high compression pistons, the later TT bikes came with slightly lower, 11 to 1 compression, in place of the very high 12 to 1 used on the first bikes.

Not a factory TT bike, in fact not a factory bike at all. Number 5 is a real-life flat track bike based on a T120TT engine with an extra 100cc of displacement thanks to a Routt 750cc big bore kit. Breathing is enhanced with the use of 34mm Mikuni carbs, a ported head, JoMo #15 cams and lightened cam wheels and rockers. The nickel plated frame is from Trackmaster, mated to forks from Ceriani and eighteen inch alloy rims from Akront. Not a museum piece, the former Larry Palmgren bike is fully restored to competition specs and ready to race.

A Sad Ending

The story of the Triumph's last years is a sad story, filled with unfulfilled promises, "what ifs" and what-might-have-beens. Development of the new three-cylinder bikes is a case in point. It wasn't that the new three cylinder bikes were bad, hell, they were good enough to win major road races against the supposedly superior Japanese bikes. It's just that everything Triumph did during the period was too little, too late.

The first triple was conceived in 1961, but getting the project off the ground and funded took a long time. According to Lindsay Brooke and Mike Duckworth, two experimental triples were built by 1966. Somehow it wasn't until 1968 that the first T150 Tridents were offered for sale. The extra years were spent designing some of the ugliest sheet metal ever seen on a motorcycle, all the work of Ogle Design. One look at the bread-box gas tanks and you have to wonder about the BSA management team that hired Ogle and approved all their work.

Aside from the styling the new triples were fast. But they were also missing a few key components. Like a disc brake, electric starter and five-speed transmission. Three items very much in evidence on the Honda 750 four-cylinder bike. The twin-cylinder bikes suffered a similar fate. Rather than deal with issues like vibration, they redesigned one of the classic designs of all time. The new 1971 bikes came with the infamous oil-bearing frame, one that raised the seat height three inches. The list of mistakes is too long and painful to list here. The good news is the fact that most of the styling blunders were eventually corrected.

The Triumph triple, as introduced up top, and as last seen on the bottom. The first T150 used a bore and stroke of 75X82mm for a total of only 724cc. Different cylinder castings with a 76mm bore brought displacement to 747cc. Sheet metal kits were shipped to the US so the ugly Ogle sheet metal could be replaced with something more Triumph-like. The 1975 Trident, the last and best, used the BSA-engine and a stretched version of the typical Triumph gas tank. The five-speed transmission and electric starter made the bike a thoroughly modern motorcycle.

Designed by Craig Vetter and very much a product of the chopper-age, the Triumph Hurricane started out as a secret BSA project. As such it uses the forward leaning BSA engine, combined with longer forks. The early demise of BSA meant that by the time the bikes were ready for production they became Triumphs. Owned by Baxter Cycle, these two are part of a production run of only 1172 bikes. The gas tank is hidden under the one-piece tank/sidecover assembly. Front hub is standard issue BSA/Triumph, though the alloy rims from Borrani are unique to the Hurricane.

The obnoxious styling (what were they smokin'?) of the early Tridents was eventually corrected, as shown by this later T150 from Baxter Cycle. Note the Triumph-style tank and softly crowned side covers. These later triples finally came with a front disc brake and five-speed transmission.

All the triples used a triplex primary chain, driving a heavy-duty clutch assembly from Borg & Beck. Because the triples were developed by basically adding a cylinder to a twin, they looked very much like a twin when viewed from the side.

As evidenced by this 1973 Tiger 750, the dread oil-bearing frame was eventually altered to bring the seat height down to a more reasonable level. This Baxter bike uses the then-standard Lockheed disc brake in front, bolted to the modern front fork with internal, not external, springs.

This Bonnie from 1979 continues the tradition with two Amal MKII carburetors mounted to a new head with parallel intake ports, all part of the 750cc engine. Owned by Dave Prichard, this late Bonneville includes the five-speed transmission, disc brakes, cast wheels, turn signals and left side shifting. The big Bonnie uses a triplex primary chain and beefier clutch assembly. Some riders feel these later bikes with extra cubes, lower compression and five-speed transmission make better daily riders than the beloved 1970 and earlier bikes.

We've all seen a Silver Jubilee or three. The Royal Wedding is similar in inspiration, though much more rare. Built to commemorate the wedding of Prince Charles to Lady Di, the bike came in both a UK and USA version. Shown here, owned by James Friddle, is a UK version with the silver frame and large tank. Most of the bike is standard-issue Bonneville, right down to the gas-tank shape, lights and specifications. The differences between this and the standard bikes lay in the chrome-finish tank and gold trim used throughout.

One of only 250 bikes built, this Royal Wedding used very few parts recognizable to old-time Triumph riders. The gas tank, sourced from Italy, is hard to identify as a Triumph tank. Bing carburetors replace the more typical Amals, and there's a button on the bars connected to an electric starter. The last Bonnevilles did come with good equipment, note the adjustable rear shock absorbers. These particular models came with a black-finish primary cover and lower fork legs. The shape of the primary cover is one of the things that didn't changed from the good old days.

Perhaps in a desperate attempt to avoid liquidation, Triumph produced a flurry of models in the last few years of operation. The TSX, or American Special, uses a chopper profile complete with a slim front fender and fat sixteen inch rear tire, all in a bid to attract new riders. Owned by Baxter Cycle, this TSX came with Bing carburetors and the unusual timing side cover installed to accommodate the electric starter and necessary drive components. Cast wheels are from Lester, brakes bear the Lockheed logo.

The best for last. This Wayne Hamilton TSS was the last to roll off the line. Conceived by Weslake Engineering as a hop-up kit for 650s, the TSS eight-valve head was used by Triumph as the centerpiece of a whole new model. More than just a four-valve head, the TSS used a new crankshaft assembly to produce a Bonneville that was fast, smooth and capable of higher revs. Produced in limited numbers in 1983, the TSS was a hell of a motorcycle, but not enough to reverse the effects of 15 years of bad management, labor strife and an enormous debt. On August 26, 1983 Triumph went into voluntary liquidation.

New/Old Triumphs

John Bloor is the man who purchased the Triumph name shortly after the closure of the plant in 1983. Definitely not a romantic, Triumph's new owner put together a plan to build new bikes in a new plant. Fresh engineers and designers combined their talents to market a range of three and four-cylinder bikes that looked more Japanese than English. The Bonneville stayed in production for a time only because the new Triumph Corporation licensed Les Harris to build T140 Bonnevilles. When Triumph closed the doors to the Meriden plant, many dealers were left with unsold stock and unpaid warranty claims. Reinventing the Bonneville and its close kin looked to the new Team Triumph like an unnecessary excuse to examine the worst parts of Triumph's recent history.

So it wasn't until the new three and four-cylinder bikes achieved success in the marketplace, and a certain amount of time had passed, that Triumph began to consider the production of nostalgia bikes built as much for their aesthetic and historical appeal as their true performance. The first retro bike was not a Bonnie but the new Thunderbird, based on a de-tuned three cylinder engine with all the right 1960s styling cues. Though the T-bird enjoyed good sales success, it really wasn't "the one."

According to author Lindsay Brooke, Triumph's engineers and stylists bought a restored 1970 Bonneville before designing the new 2001 Bonneville. Thus the legacy of Edward Turner's Speed Twin continues. All of which goes to show that some designs, the really good ones, just won't go away.

Parked above the bike it is patterned after, this new Bonneville shares the field with a 1970 Bonnie (which should have the throttle cables routed through the headlight ear). To develop the new Bonneville, Triumph developed a new 800cc overhead cam twin, designed to look as much as possible like a certain 650cc pushrod twin developed more than sixty years earlier. The Bonneville frame too is also new from the ground up, and uses the engine as a semi-stressed member. To stay true to the bike's roots, twin rear shocks are part of the kit, along with laced wire wheels.

Like any successful model, the new Bonneville is the source of various spin-offs. The Golden Jubilee commemorates the fiftieth year of Queen Elizabeth's rule, while borrowing styling cues from the Silver Jubilee, designed to commemorate twenty-five years of her reign. Monday-morning quarterbacks are quick to criticize the new bike's taillight, heavy rear fender and "kinky" exhaust pipe. Perhaps the new Bonneville and its various spin offs should be viewed as spiritual successors, rather than perfect clones, of the original bike.

BSA and Triumph were arch rivals in the old days. Yet, it is a BSA model that served as inspiration for the biggest motorcycle ever built with a Triumph badge on the tank. Displacing a full 2300cc the three-cylinder Rocket III uses a 150X17 front tire, and a 240X16 in the rear, to handle 147 foot pounds of torque. Where the old bikes used chain drive and a four-speed transmission, the new Rocket machine relies on a shaft drive and five-speed. Triple disc brakes are used to slow the considerable momentum of a 700 pound motorcycle.

Wolfgang Books On The Web

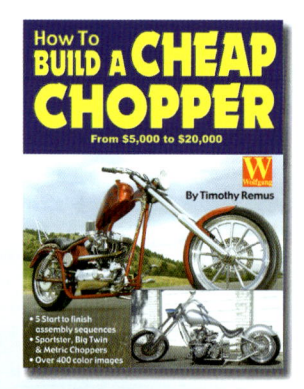

How to Build A Cheap Chopper

Choppers don't have to cost $30,000.00. In fact, a chopper built at home can be had for as little as $5,000.00. Watch the construction of 4 inexpensive choppers with complete start-to-finish photo sequences. Least expensive are metric choppers, based on a 1970s vintage Japanese four-cylinder drivetrains installed in an hardtail frame. Next up are three bikes built using Buell/Sportster drivetrains. The fact is a complete used Buell or Sportster is an inexpensive motorcycle – and comes with wheels and tires, transmission, brakes and all the rest. Just add a hardtail frame and accessories to suit. Most expensive is bike number 4. This big-twin chopper uses a RevTech drivetrain set in a Rolling Thunder frame.
Written by Tim Remus. Shot in the shops of Tom Summers, Brian Klock, Motorcycle Works, Redneck Engineering and Dave Perewitz to illustrate the start-to-finish construction of these 4 bikes.

Eleven Chapters   144 Pages $24.95   Over 400 photos-100% color

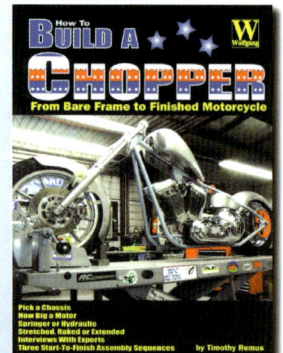

How to Build A Chopper

Designed to help you build your own chopper, this book covers History, Frames, Chassis Components, Wheels and Tires, Engine Options, Drivetrains, Wiring, Sheet Metal and Hardware. Included are assembly sequences from the Arlen Ness, Donnie Smith and American Thunder shops. Your best first step! Order today.

Choppers are back! Learn from the best how to build yours. 12 chapters cover:
- Use of Evo, TC, Shovel, Pan or Knucklehead engines
- Frame and running gear choices
- Design decisions - short and stubby or long and radical?
- Four, five or six-speed trannies

Twelve Chapters   144 Pages $24.95   Over 300 photos-over 50% color

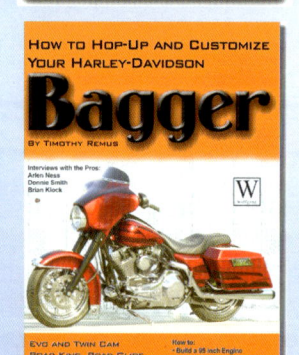

Hop-Up & Customize Your H-D Bagger

Baggers don't have to be slow, and they don't have to look like every other Dresser in the parking lot. Take your Bagger from slow to show with a few more cubic inches, a little paint and some well placed accessories. Whether you're looking for additional power or more visual pizazz, the answers and ideas you need are contained in this new book from Tim Remus.

Follow the project bike from start to finish, including a complete dyno test and remapping of the fuel injections. Includes two 95 inch engine make overs.
How to:
- Pick the best accessories for the best value
- Install a lowering kit
- Do custom paint on a budget
- Create a unique design for your bike

Eight Chapters   144 Pages $24.95   Over 400 photos   100% color

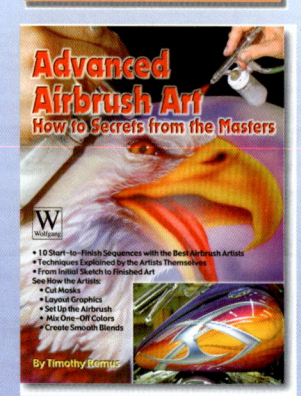

Advanced Airbrush Art

Like a video done with still photography, this book is made up entirely of photo sequences that illustrate each small step in the creation of an airbrushed masterpiece. Interviews explain each artist's preference for paint and equipment, and secrets learned over decades of painting. This is a great book for anyone who takes their airbrushing seriously and wants to learn more.

Learn How To:
- Pick an Airbrush
- Design a Graphic
- Utilize the Computer
- Choose the Best Colors

Ten Chapters   144 Pages $24.95   Over 500 photos, 100% color

Visit us at: www.wolfpub.com

ADVANCED CUSTOM PAINTING TECHNIQUES

When it comes to custom painting, there is one name better known than all the others, and that name is Jon Kosmoski. Whether the project in your shop rides on two wheels or four, whether you're trying to do a simple kandy job or complex graphics, this how-to book from Jon Kosmoski is sure to answer your questions. Chapters one through three cover Shop Equipment, Gun Control and Paint Materials. Chapters four through seven get to the heart of the matter with complete start-to-finish painting sequences.

- Shop set up
- Gun Control
- Use of new paint materials
- 4 start-to-finish sequences
- Two wheels or four
- Simple or complex
- Kandy & Klear

Seven Chapters   144 Pages $24.95   Over 350 photos, 100% color

HOW TO: CUSTOM PAINT & GRAPHICS

A joint effort of the master of custom painting, Jon Kosmoski and Tim Remus, this is the book for anyone who wants to try their hand at dressing up their street rod, truck or motorcycle with lettering, flames or exotic graphics. A great companion to Kustom Painting Secrets. Includes both hand lettering, pinstriping and extensive airbrush work.

7 chapters include:
- Shop tools and equipment
- Paint and materials
- Letter & pinstripe by hand
- Design and tapeouts
- Airbrushing
- Hands-on, Flames and signs
- Hands-on, Graphics

Seven Chapters   144 Pages $24.95   Over 250 photos, 50% in color

ULTIMATE SHEET METAL FABRICATION

In an age when most products are made by the thousands, many yearn for the one-of-kind metal creation. Whether you're building or restoring a car, motorcycle, airplane or (you get the idea), you'll find the information you need to custom build your own parts from steel or aluminum. Includes both simple and complex projects made with both hand and power tools. Also, welding sheet metal with gas or tig.

11 chapters include:
- Layout a project
- Pick the right material
- Shrinkers & stretchers
- English wheel
- Make & use simple tooling
- Weld aluminum or steel
- Use hand and power tools

Eleven Chapters   144 Pages $19.95   Over 350 photos

ADVANCED SHEET METAL FABRICATION

Advanced Sheet Metal Fabrication Techniques is a photo-intensive how-to book. See Craig Naff build a Rolls Royce fender, Rob Roehl create a motorcycle gas tank, Ron Covell form part of a quarter midget body and Fay Butler shape an aircraft wheel fairing. Methods and tools include English wheel, power hammer, shrinkers and stretchers, and of course the hammer and dolly.

- Sequences in aluminum and steel
- Multi-piece projects
- Start to finish sequences
- From building the buck to shaping the steel
- Includes interviews with the metal shapers
- Automotive, motorcycle and aircraft

7 Chapters   144 Pages $24.95   144 pages, over 300 photos - 60% color

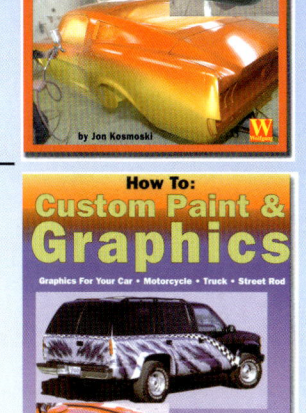

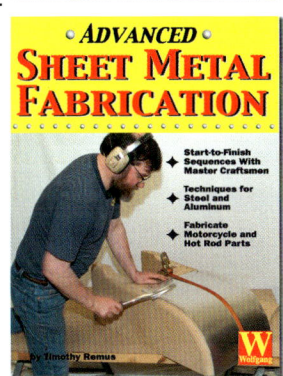

Baxter Cycle
PO Box 85
Marne, IA 51552
712.781.2351
Fax: 712.781.2355
e-mail: bikes@baxtercycle.com
web page: www.baxtercycle.com

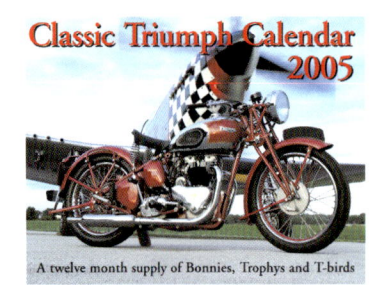

In addition to their vast inventory of British parts and motorcycles, Baxter Cycle stocks T-shirts, hats and the most current version of the Classic Triumph Calendar.

144